Making Equipment-Replacement Decisions

Lawrence C. Hackamack

AMA

Making Equipment-Replacement Decisions

Making Equipment-Replacement Decisions

Lawrence C. Hackamack

American Management Association, Inc.

This book has been distributed without charge to AMA members enrolled in the Manufacturing Division.

Standard book number: 8144–2131–8
Library of Congress catalog card number: 77–99734

Acknowledgments

The author wishes to thank the following people for their generous assistance in the preparation of this book: Andy Ashburn, Editor of *American Machinist;* Les Barnes of Northern Illinois University; Ira Doell of General Electric Company, Schenectady, New York; George Daniels of Rock Island Arsenal, Rock Island, Illinois; and Ted Koyzis of Marketron Associates, Wheaton, Illinois. A special debt of gratitude is owed the author's family for its support and to Northern Illinois University for its cooperation.

Contents

Making Equipment-Replacement Decisions

INDUSTRY needs a sensible approach to investment analysis more than it needs a sophisticated one. As a matter of fact, the crudest tool that will give the right answer is the one that should be used. Only as investments become more and more marginal, or as the value goes up, does it become really worthwhile to refine the approach.

The same analysis should not be used for every prospective equipment investment; each study should be tailored to meet the demands of the particular circumstances. A real effort needs to be applied to get, not greater precision, but realistic and uniform estimates of such basic factors as the net investment, savings, added revenue, intangible benefits, and projected life. If this has been done properly, making the decision to invest becomes an easy matter.

As simple as this approach may be, getting management to use it is another matter. As Charles Kettering once remarked when referring to progress in physical technology: "You can communicate a message around the world in seven seconds, but for a simple idea to penetrate a quarter inch of human skull may require generations."

Since a company is seldom able to purchase everything that it requires in the way of equipment, it must judiciously invest its capital where it can accomplish planned objectives designed to produce the best profits. Good equipment is the key to profitability, but obviously all good equipment is not equally profitable. Therefore, to make intelligent decisions, management needs a program that will reveal many worthwhile projects for evaluation and will provide a sound, objective method for selecting the most advantageous solution to a problem. The purpose of the following related material is to develop performance guidelines for an effective equipment program, to review the major justification methods, and to consider the uses of cost information.

Setting Up a Replacement Program

The average age of machine tools in the United States —more than ten years—has been well publicized, but this statistic has had little noticeable impact on management. The problem of equipment obsolescence has been a source of concern to those who would like to sell more machine tools; to those who must prepare for industrial mobilization; and to those who are aware of increased foreign competition. It was not until recently, however, that equipment obsolescence became the concern of those whose responsibility it is to operate and maintain such equipment. Now that foreign competition has become more intense and the Department of Defense has become more insistent that items be manufactured with up-to-date facilities, plant modernization is finally receiving positive attention. Most managements, having limited funds for replacing outmoded equipment, seek programs that will effectively disclose good replacement opportunities and provide for the best use of their capital.

An effective modernization program should perform the following functions:

1. Alert management to profitable replacement opportunities.
2. Collect the data that are pertinent to the decision.
3. Estimate the profitability of the proposal.
4. Determine the need for funds and the funds available.
5. Select the best investment opportunities for the available funds.

Upon learning of a new machine, it is possible for the owner of a small business to complete the preceding process mentally and make a decision in a few minutes; however, a major corporation might have to use a more formal, sophisticated procedure. The results in either case can be good or bad, depending upon the quality and the level of skill and thought applied. An organized replacement program therefore, requires the assignment of competent persons for the performance of each step of the program.

Once these persons have been assigned the specific responsibility of indicating the need for replacement or modernization projects, how do they actually accomplish the task? In addition to learning about the newest equipment on the market, they must be thoroughly familiar with their own production methods. Ideally, too, they will have imagination. They will not merely seek newer models of old equipment but will consider redesigned products and take advantage of completely new processes. Also, they will constantly survey their operation for problems and potential problems arising from obsolete or obsolescent equipment. Finally, they will learn to recognize the clues that will lead them to replacement possibilities.

Obsolescence is not always a sure indicator of the usability of a machine. A ten-year-old machine in some applications may be completely adequate. If the utilization of the machine is infrequent, or if production requirements do not demand the use of higher feeds and speeds, the purchase of a newer model may be unjustified. Although constant use of a machine is not a sure sign that the machine should be replaced, high utilization could be a reflection of low efficiency. Indeed, this may be an occasion for investigation of the latest version with improved feeds and speeds or easier setups. The replacement of an obsolete machine may not only reduce costs, it may also increase the capacity of an entire production system.

If a large part of a machine's nonproductive time involves preparing a setup for short runs, use of a numerically controlled machine might be considered. Many men running job-shop operations are unaware of the fact that numerical control offers a greater economic advantage to them than to long-run production shops.

A heavy concentration of manpower to operate a machine is another indication of the need for replacement. The fact that a machine requires one or more helpers to perform simple handling or control tasks should trigger a quick investigation of current models of that machine (or alternate methods). Newer models might perform the work of these helpers automatically, while at the same time provide savings. Savings in direct labor, even if they are too small to justify a capital expenditure by themselves, are usually an indication that full-scale evaluation is in order, since other savings and advantages may result from the study.

In a survey conducted by the author for the Federal Reserve Board of Boston, Massachusetts, in the fall of 1960 and published by Boston University in *Business Review,* interviews were conducted with budgetary officers faced with the decision of whether to purchase a new piece of equipment. On the basis of these interviews, it was decided that certain factors should influence a decision in favor of replacement: (1) the desirability of having an unusual piece of equipment that would advertise itself; (2) an absolute need for new equipment because present equipment was no longer capable of repair; (3) techniques or formulas established by the Machinery and Allied Products Institute (MAPI), (4) the service given by the manufacturer or the distributor; (5) worker morale and the accident potential; (6) the difference between management and shop evaluation; and (7) the quality of the finished product. Some of the firms in the study maintained adequate budgetary devices to provide for replacements, while others did not. There were few, if any, attempts made to develop cash reserves.

Often, inexperienced line or staff officers in middle management tend to submit numerous suggestions regarding equipment replacement in order to achieve what they consider to be the desired effect of maintaining or improving the current level of operations in each of their departments. While these officers are able to supply information concerning the needs of companies for new equipment, ultimately they come up against the same problem—"to buy or not to buy." What contributes to making this such a difficult decision is that, in very simple terms, the amount of dollars to be used for replacement is restricted by capital budgeting concepts—concepts that are extremely conservative and restrictive. These concepts are based on the premise that either the value rendered by a new piece of equipment must result in a return on the investment dollar or there must be a realized saving great enough to justify its installation. By way of illustration, it is found that some of the large corporations adhere to the philosophy that a two-and-a-half year return is all that should be permitted. No purchase beyond this rate of return will be made. This conservative approach means that either the managers are not aware of the restrictive nature of the policies they have established, or they do not realize the changes that have occurred in the productivity of the equipment currently available.

All too often, staff managers do indicate that they realize the need for equipment replacement. However, it is management at the higher levels that needs to re-examine this need within the context of the overall operation of the firm. The restrictions placed upon managers before they consider a particular piece of equipment must, by necessity, conflict with the firm's advertising and other required expenditures. Generally speaking, there is a lack of understanding by management of the impact that new equipment will have on the firm in the future.

It has been said by some that it is too bad that American production facilities were not destroyed as Germany's were during World War II. Then American businesses could have started over with completely modern equipment. Any top-level manager who fails to realize that by basing his decision-making evaluation pertaining to equipment matters upon the belief that such investments must yield a return on investment identical to the amount that might be realized in investments in other areas is forcing his company into a serious situation, because he is not considering the basic change in productivity that equipment-replacement expenditures can produce. Thus a careful value analysis should be made before deciding for or against equipment replacement.

Equipment obsolescence and replacement are fields that cause manufacturers much anxiety. Any business adopting long-term goals cannot ignore the continual technological advances which affect production efficiency. In a free, competitive society a firm with production costs that are above those of a competitor does not have much chance for survival. The Society for the Advancement of Management has made popular the following industrial maxim: "No war, no strike, no depression can completely destroy an established business or its profits, as new and better methods, equipment, and materials in the hands of an enlightened competitor."[1]

The main problem concerning equipment replacement is that an old machine will probably still function as effectively as it did when it was new, but the fact remains that technological advances in productivity may have outmoded it. The cost of production on old machines increases or remains the same; however, competitors may have incorporated the benefits of lowered production costs through the utilization of new equipment with the latest technology.

Some of the important reasons for replacing equipment are as follows: (1) to produce better products; (2) to remain in a competitive position; (3) to increase output; (4) to provide stockholders with a greater return on investment; (5) to cut maintenance costs; (6) to reduce the waste and scrap factor; (7) to lower direct labor cost; and (8) to reduce work in process inventories.

There are also noneconomic considerations that a company should incorporate into its equipment replacement policy, such as: (1) improving the safety of an operation; (2) reducing the human-error factor; (3) increasing the efficiency of operations; and (4) increasing the adaptability of equipment.

The following reasons for replacing equipment also warrant careful consideration:

Basically, there are four situations that prompt the selection of equipment: (1) the substitution of a machine for a manual operation; (2) the relative desirability of two or three machines proposed for the same operation; (3) the replacement of an old machine with a new and improved one; (4) the relative desirability of replacing old equipment with identical new equipment.[2]

The first, crucial step toward the efficient handling of replacement programs is obtaining recognition by top management that replacement problems do exist and that they should be handled on a systematic basis. With such a systematic approach careful planning based on complete data will greatly reduce the number of mistakes made in the purchase of equipment. The advantages of systematic replacements are as follows: reduced maintenance and production costs, lower losses from scrap and rework, and less lost time, as well as an increase in the enthusiasm and morale of employees.

As Manley H. Jones put it: "Despite the fact that we have the most advanced industrial technology in the world, there is still a large amount of waste and inefficiency in our plants. The most progressive manufacturing firms gradually eliminate such inefficiencies (resulting in their replacement by new and unforeseen inefficiencies), while at firms with less alert and capable top management, waste and inefficiency continue with a minimum amount of concern or interference."[3] In other words, firms with less capable top managements continue to incur costs caused by inefficiency.

The effective top-level manager should ask himself the following questions: Am I confident that the capital-equipment proposals I accept are more attractive than those I reject? Am I satisfied that none of the proposals that I have rejected would have been profitable if I had implemented them? There are several techniques available to the top-level manager to aid him in determining the answers to these questions. We shall examine several of these techniques and compare them with a view to judging the relative merits of each.

Productivity Criteria Quotient

One of the most recent techniques developed for equipment evaluation is the Productivity Criteria Quotient (PCQ). The purpose of PCQ, as it was devised by the author in 1960, is to determine increases in productivity and process capability from technological changes made in machines. Most methods of equipment replacement place a time element on the usefulness of machines. In many cases the technique of assigning a period to the usefulness of a machine can be detrimental to the best interests of the company. A good example of this would be designating the economic life of a machine as eight years, when its useful life may only be four years because of technological advances in the equipment. The reverse may also be true—that is, an old machine may be considered obsolete, while a new machine may perform only slightly better than the existing piece of equipment. Equipment life must be determined on the basis of the significant changes that have been made in machine tools.

The PCQ technique evaluates the machine itself and does not try to evaluate the constantly changing output per man-hour. Profit comes from the marketplace; productivity comes from the plant. Profitability comes from product acceptance and the effective use of capacity to meet the market need. Therefore, manufacturing-equipment selection must be based on profitable productivity—that is, the best combination of process productivity and product marketability.

The purpose of the PCQ method is to supply management with knowledge concerning the changes in machine tools that have taken place from 1940 to date, thereby permitting a better selection of machine tools to meet the demands of today. Very often these changes have a direct bearing on a machine's overall productivity. A survey of 61 companies was made to determine what machine improvements were actually made. Changes made in each machine type were recorded by year as were recommendations for each category's replacement. It was discovered that using outmoded equipment often results in excessive costs and greatly increased production time. Clearly, a company must have modern, up-to-date machines in order to remain competitive.

Thirteen types of machines were selected as representative of the various trends in machine modernization: horizontal boring and milling, vertical boring and milling, vertical jig boring, radial drills, sensitive drills, cylindrical grinders, surface grinders, engine lathes, turret lathes, horizontal milling, vertical milling, contour saws, and cutoff saws.

Each of the 13 machine types was studied, and certain criteria changes were established for each. These criteria were evaluated on the basis of data obtained from a questionnaire mailed to 67 companies. Only when a machine-tool company produced a unit that satisfied the criteria, was it possible to have the change recorded. Thus the number of changes do not necessarily reflect additions to a machine, but rather basic, substantial improvements. The study does not record the introduction of new models by the companies; it is a documentation of the improvements that have been made on the actual tools.

Since 1955 rapid changes in technology have caused equipment to become outmoded at a faster rate than in any other period. Consequently, major replacement must be practiced at an increasing rate to maintain the high capability that is characteristic of today's manufacturing. This book will not attempt to discuss the individual machine categories, but will treat the general methods of PCQ. Exhibit 1 shows the number of changes in the 13 machine categories that have taken place since 1940.

An examination of Exhibit 1 reveals that during World War II there were relatively few changes in machine design. The first big increase in design changes came in 1947, followed by significant increases in 1955 and 1960; interestingly enough, these increases coincided with the major machine-tool shows. In other words, the rising innovation curve reflects a corresponding increase in research and development.

EXHIBIT 1

PCQ Change Table (1940-1967)

YEAR	*BJ*	*BV*	*DR*	*DS*	*ED*	*GC*	*GS*	*LE*	*LT*	*MB*	*MM*	*SC*	*ST*
40	0	140	10	4	0	0	23	10	8	6	6	0	20
41	0	140	10	4	0	0	28	25	8	20	6	4	27
42	0	140	10	4	0	0	28	25	10	30	12	4	27
43	0	140	10	6	0	0	30	25	20	32	12	4	37
44	0	140	10	6	0	0	30	25	25	62	28	4	37
45	7	165	10	6	0	0	33	36	53	64	28	4	37
46	37	165	10	6	0	5	35	59	61	74	32	12	40
47	67	165	32	6	0	75	40	109	123	86	70	12	40
48	67	165	32	6	0	75	43	130	128	90	70	12	40
49	94	175	32	6	0	135	43	146	166	120	74	12	60
50	94	190	32	8	0	145	50	170	188	172	80	12	63
51	94	195	32	8	0	145	58	173	208	216	80	12	63
52	94	300	32	10	0	145	63	178	213	220	80	12	70
53	94	330	32	10	0	150	68	184	225	222	80	26	77
54	94	330	32	10	0	155	78	189	253	242	80	32	77
55	124	405	67	22	10	180	80	280	311	276	150	32	77
56	137	435	75	26	13	202	80	286	343	300	158	50	77
57	170	470	83	32	29	205	107	287	343	350	168	64	77
58	173	505	103	48	29	227	115	290	373	364	168	72	90
59	196	630	115	52	29	247	118	309	383	382	174	86	90
60	253	640	165	70	44	272	152	391	465	472	198	96	93
61	253	660	165	70	44	272	152	391	478	484	200	99	93
62	253	675	165	78	46	272	154	395	478	496	200	99	93
63	253	695	165	83	54	272	167	395	490	496	200	102	93
64	270	695	165	96	62	272	184	408	505	504	204	105	103
65	287	710	173	103	70	282	198	417	518	516	211	122	126
66	295	716	173	108	76	287	206	423	524	516	217	137	131
67	295	716	183	121	109	287	226	426	526	531	227	137	136

BJ – Jig Bore
BV – Vertical Boring
DR – Radial Drill
DS – Sensitive Drill
ED – Electric Discharge
GC – Cylindrical Grinder
GS – Surface Grinder
LE – Engine Lathe
LT – Turret Lathe
MB – Horizontal Boring Mill
MM – Milling Machine
SC – Cutoff Saw
ST – Contour Saw

In the October 7, 1968, issue of *American Machinist* the machine categories mentioned earlier have been plotted on charts (see Appendix) that show changes in critical criteria. Each type of machine has certain critical elements of design and construction that have a direct bearing on the ability of the machine to do its job.

Each of these elements (called critical criteria) is given a weight of 1, 2, or 3. Actual changes in the design or construction of lines relating to these machines are studied. Each improvement that affects one of the critical criteria is tabulated, while changes that do not affect the critical criteria are ignored. Each table shows the total changes in critical criteria for each year from 1950 to 1967. The PCQ value is obtained by taking each year's total of critical criteria changes and dividing the number of machine lines studied so that all tables will be based on one average line of machines. This value is then multiplied by ten to make the number more convenient to work with, and the result is added to the result of the previous year to accumulate the results from year to year.

The following example shows how PCQ can be used: Looking at the table of the horizontal boring mills in the Appendix, we see that in 1950 the PCQ was 172; in 1955 it was 276; in other words, between 1950 and 1955 these machines improved by 104 (276 − 172). Comparing 1955 and 1960, the PCQ's are 276 and 472 respectively. The PCQ gain during this period is 196—almost twice the figure for the previous five-year period.

To carry this example a step further, assume that a new machine which costs $15,000 is being considered to replace a machine whose present value is $5,000. The cost to replace the machine would be $10,000, which would be paid for in five years. To determine what it costs to retain the old equipment over the new equipment, multiply the difference in cost by the change in PCQ value; then divide by one hundred. Consider the following example:

Cost of new machine	$15,000
Present salvage value	$ 5,000
PCQ value of 5-year-old machine	172
PCQ value of new machine	276
Cost per year not to replace new machine	

$$= (\$15{,}000 - \$5{,}000) \times \frac{(276 - 172)}{100}$$

$$= \$10{,}000 \times 1.04 = \$10{,}400$$

If the PCQ values were 210 and 260, then the problem would be solved as follows:

Cost per year not to replace new machine

$$= (\$15{,}000 - \$5{,}000) \times \frac{(260 - 210)}{100}$$

$$= \$10{,}000 \times .50 = \$5{,}000$$

PCQ ratings do not provide an exact measure of gain in productivity; nor are they a substitute for a detailed analysis of a proposed new machine. They do, however, indicate the relative rate of change in a machine type at different times.

Payoff-Period Method

A simple problem will be used to illustrate solutions by the payoff-period method (as well as the other methods considered in this book). The problem is whether a new machine costing $20,000 should replace an obsolete machine whose present book value is $6,000. The problem has been simplified so that the service and tax lives are the same, and there is no salvage value. Straight-line depreciation is used, as well as a 50 percent income tax rate; and the annual costs remain constant. The problem statement is as follows:

PROBLEM STATEMENT

	Present Machine	*Proposed Machine*
Installed cost	$10,000	$20,000
Present book value	$ 6,000	—
Present salvage value	$ 5,000	—
Salvage value at end of life	—	$ 1,000
Average annual operating cost	$ 8,000	$ 5,000
Service life	1 year	8 years
Annual depreciation	$ 1,000	$ 2,500
Income tax rate	50%	50%

The payoff period varies from one company to another. Some companies require 100 percent capital recovery, plus a minimum attractive return within a certain period. Other companies require merely 100 percent capital recovery with no return; a shorter period compensates for no rate of return. Various studies[4] indicate that most companies require a payoff on machinery of five years or less and that a substantial number of companies require three years or less. A two- or three-year payoff would be appropriate for special machines that have a short life. If a short write-off period is used indiscriminately to purchase all machinery, then the extremely short payoff requirement would interfere with many sound proposals that might be economical. The reason many owners use a writeoff is that the proposed new asset will decline rapidly in the early years because of the availability of new equipment or changes in requirements. The basis for a short payoff requirement is past experience with obsolescence and inadequacy of similar assets, as well as the belief that history will repeat itself. This often leads to much more drastic requirements than are really justified. Excessively short payoff requirements, such as two or three years for assets that are capable of serving ten years or more, are almost never justified, except for special-purpose assets when the continuation of the service need is in doubt. Such excessive safety factors are undesirable in economic replacement studies, because they tend to perpetuate obsolete and uneconomical plants and equipment. This technique does not consider when incomes are earned, nor do they consider what happens after the payout period. Such a method can be dangerous when comparing alternatives with unequal machine life, and can lead to accepting investments that are actually unattractive.

An example of this procedure is the following: Consider two machines that cost \$20,000 and \$22,000 respectively, with each having a return of \$10,000 per year. The two-year payoff criteria would accept the first machine and reject the second. Suppose the first machine had a life of two years, and the second had a life of five years. The two-year machine would return nothing on investment, but the second machine would return a profit of \$28,000 for the five years. A zero rate of return would be accepted over a possible 13 percent return.

Following are a few base methods used in determining the payoff period:

$$\frac{\text{Gross investment}}{\text{Annual savings}} = \frac{\$20{,}000}{\$8{,}000 - \$5{,}000} = 6.7 \text{ years}$$

$$\frac{\text{Net investment}}{\text{Annual savings}} = \frac{\$20{,}000 - \$5{,}000}{\$8{,}000 - \$5{,}000} = 5 \text{ years}$$

$$\frac{\text{Net investment}}{\text{Annual savings after taxes}} = \frac{\$20{,}000 - \$5{,}000}{\$3{,}000 - (\$3{,}000 - \$2{,}000)(.50)} = 6 \text{ years}$$

$$\frac{\text{Net investment}}{\text{Annual savings after taxes and depreciation}} = \frac{\$20{,}000 - \$5{,}000}{\$3{,}000 - (\$1{,}000)(.50) - \$2{,}000} = 30 \text{ years}$$

We assume that a new machine has been purchased for \$20,000 to replace a machine with a salvage value of \$5,000. This means that there has been a net investment of \$15,000. In a given year the new machine costs \$5,000 to operate, and the old machine costs \$8,000 to operate, which yields a saving of \$3,000 per year; the machine will depreciate at the rate of \$2,000 per year.

Rate-of-Return Method

Using the rate-of-return concept for investment in new equipment involves setting a minimum rate of return, usually established by top management, when proposing a project. Although this method has limitations when used for evaluating equipment replacements, its principal value is in ranking replacement projects. The rate-of-return approach can be the common denominator for ranking replacement projects and selecting those with the biggest payoff. Like the payoff approach, the rate-of-return calculations are influenced by the manner in which the investment and savings are stated.

The rate-of-return method is designed to find the interest rate at which the two money time series are equal—that is, when the payback and rate of return yield equal results. The interest rate represents the return on the extra investment required for one of a series. If the return on the extra investment is satisfactory, as determined by company policy, then a decision to make the extra investment is made.

In the previously mentioned example a new machine was purchased to replace an old machine that had a salvage value of \$5,000. The figures from that example will be used to illustrate the rate-of-return method as follows:

$$\frac{\text{Annual savings}}{\text{Gross investment}} = \frac{\$8{,}000 - \$5{,}000}{\$20{,}000} = 15\%$$

$$\frac{\text{Annual savings}}{\text{Net investment}} = \frac{\$8{,}000 - \$5{,}000}{\$20{,}000 - \$5{,}000} = 20\%$$

$$\frac{\text{Annual savings after taxes}}{\text{Net investment}} = \frac{\$3{,}000 - (\$3{,}000 - \$2{,}000) \times (.50)}{\$20{,}000 - \$5{,}000} = 16.7\%$$

$$\frac{\text{Annual savings after taxes and depreciation}}{\text{Net investment}} = \frac{\$3{,}000 - (\$1{,}000) \times (.50) - \$2{,}000}{\$20{,}000 - \$5{,}000} = 3.3\%$$

$$\frac{\text{Average annual savings}}{\text{Average book values}} = \frac{\$24{,}000/8}{(\$20{,}000 - \$1{,}000/2)} = 31.7\%$$

$$\frac{\text{Average annual savings}}{\text{Book value}} = \frac{\$24{,}000/8}{\$20{,}000} = 15.8\%$$

This method answers a very basic question: What annual interest rate must be applied to a deposit or investment of \$15,000, and subsequent balances on that deposit, if one expects to withdraw \$3,000 at the end of each year, including principal and interest, particularly after the interest has been applied each year? Further, at the end of eight years would there be a zero balance? The solution to this problem would be calculated as follows:

$$\$15{,}000 = \frac{\$3{,}000}{(1+r)^1} + \frac{\$3{,}000}{(1+r)^2} \ldots + \frac{\$3{,}000}{(1+r)^8}$$

$$\$15{,}000 = \$3{,}000 \left[\frac{1}{(1+r)^1} + \frac{1}{(1+r)^2} \ldots + \frac{1}{(1+r)^8} \right]$$

$$\frac{\$15{,}000}{\$3{,}000} = \frac{1}{(1+r)^1} + \frac{1}{(1+r)^2} \ldots + \frac{1}{(1+r)^8}$$

$$5 = \frac{1}{(1+r)^1} + \frac{1}{(1+r)^2} \ldots + \frac{1}{(1+r)^8}$$

Consult the present-value table (\$1 per year for 8 years) and find the rate where the percent is equal to 5. The annual interest rate (r) in this example will equal 11.6 percent.

Present-Worth Method

The concept of present worth, also known as discounted cash flow, has many uses in evaluating new equipment. However, the principal objection to its use is that the results are difficult to interpret, because managers are accustomed to thinking in terms of annual cost and find it difficult to realize that a small change in the interest rate might dictate another course of action.

Consider these alternatives[5]: *A* spends \$3,000 annually in perpetuity; *B* requires \$50,000 with no annual expenditures (the capitalized cost is derived by dividing the annual costs by the interest rate). The results would be as follows:

Interest Rate	*Capitalized Cost of A*	*Capitalized Cost of B*	*Result*
3%	$100,000	$50,000	Advantage of $50,000 for *B*
6%	50,000	50,000	No advantage for either
8%	37,500	50,000	Advantage of $12,500 for *A*

At the 3 percent interest rate, *B* clearly has the advantage, while at the rate of 8 percent the advantage is reversed. Many analysts who use the present-worth method use the base cost of the borrowed money required to justify an investment, rather than the rate-of-return method. The present-worth method consists of reducing all the receipts and expenditures for each alternative to a present-worth basis. When using the lowest worth as the better investment, it is important to compare equal services. For instance, if one alternative furnishes a service for five years and another alternative furnishes it for ten years, the comparison must be made on the basis of the lowest common multiple—ten years.

Following is an example which shows the present worth as applied to the sample problem. The factors were obtained from a table that lists present-worth factors, where both the old and new machines are assumed to be in operation for eight years. In the beginning the new machine would have a market or residual value of $5,000, but would have no value in eight years. Furthermore, the time value of the money to the firm is assumed to be 10 percent per year. The new machine has an advantage of $3,660 over the present equipment.

Old Machine

Present worth of an expenditure for operating expenses of $8,000 for 8 years:

$$\$8{,}000 \times 5.335 = \$42{,}680$$

New Machine

Present worth of net expenditure for new machine:

$$\$20{,}000 - \$5{,}000 = \$15{,}000$$

Present worth of an expenditure for operating expenses of $5,000 per year for 8 years:

$$\$5{,}000 \times 5.335 = \$26{,}675$$

Present worth of $1,000 received 8 years from now:

$$\$1{,}000 \times 4.665 = \$4{,}665$$

$$\textit{Total} = \$46{,}340$$

Annual-Cost Method

As its name implies, the annual-cost method—also called the capital-recovery method—is one in which the annual cost for obtaining a service from different sources is developed. Essentially, the annual cost consists of the cost of getting the original investment back, the cost of getting a return on that investment, and the cost of operating under that particular alternative. When a salvage value is involved, the interest that could have been earned on that amount each year, had it been used otherwise, is also an annual cost of capital.

Two steps are involved when figuring the annual-cost method:

1. Capital recovery—the amount that will just repay the initial investment at a stipulated interest over the economic life of the equipment.

   $$\text{ACCR} = (\text{P} - \text{L}) \times (\text{capital-recovery factor}) + \text{L} \times \text{i}$$

   Where: ACCR = annual cost of capital recovery
   P = first cost
   L = prospective salvage value at end of life
   i = interest rate in decimals

   Formula for the capital-recovery factor (CRF):

   $$\text{CRF} = \frac{\text{i}(1+\text{i})^{\text{n}}}{(1+\text{i})^{\text{n}-1}}$$

   n = life in years

2. Annual operating cost—all relevant operating costs converted to an annual basis and then simply added to the annual cost of capital recovery. Operating costs are those normally considered in any cost analysis—such as labor, tooling, maintenance, and scrap. In most cases the average of the series of annual disbursements is used as an approximation.

This method is used to compare mutually exclusive alternatives; it is practical for use in estimating the economic life of existing or proposed equipment. The annual-cost method forces the individual to recognize that the capital cost of an existing asset is its market value, not its book value. Furthermore, the annual-cost method does not require the same termination date for each alternative, and so is well suited to replacement studies.

The following example shows how the sample machine problem is applied to the annual-cost method:

Old Machine

$$\begin{aligned} \text{ACCR} &= (\text{P} - \text{L}) \times (\text{CRF}) + \text{L} \times \text{i} \\ &= (\$5{,}000 - 0) \times (.18744) + (0) \times (.10) \\ &= \$937.20 \end{aligned}$$

Operating cost = ($8,000)
Total average annual cost = $8,937.20

Proposed Machine

$$\begin{aligned} \text{ACCR} &= (\text{P} - \text{L}) \times (\text{CRF}) + \text{L} \times \text{i} \\ &= (\$20{,}000 - \$1{,}000) \times (.18744) + \\ &\quad (\$1{,}000) \times (.10) \\ &= (\$19{,}000) \times (.18744) + 100 = \$3{,}761 \end{aligned}$$

Operating cost = $5,000
Total average annual cost = $8,761

The results indicate that the proposed machine has a lower annual cost than the present machine. The proposed machine yields $176 before taxes beyond the 10 percent rate of return.

The MAPI Method

Another approach to capital investment is the MAPI method, developed by George Terborgh for the Machinery and Allied Products Institute (MAPI). Since its introduction, the MAPI method has been modified several times. In this book a modified version of the concept will be presented, and the presentation has been simplified so that it can be easily understood. The theory of machine-tool replacement will not be treated in detail, so for further study *Dynamic Equipment Policy* and *MAPI Replacement Manual* by George Terborgh are recommended.

The replacement analysis is concerned with the problem of when to re-equip. It is this concern with timing that brings the problem of equipment replacement closer to a practical solution. The advantages of using the MAPI analysis are as follows:

1. It gives a definite answer to the question of when to replace a piece of equipment economically.
2. It provides a usable measure of the cost of not replacing, a very important item that other replacement techniques omit. It tells management the cost of deferring replacement, which is expressed as a minimum cost for the next year.
3. It furnishes a good tool for budgeting expenditures. Most companies cannot afford to purchase all new machines for replacement programs; this method, however, reveals the most serious cost offenders.
4. It is applicable to a single unit or a combination of machines in line for replacement or expansion.
5. It makes allowances for future deterioration and obsolescence expected on the new equipment—factors that are expressed in terms of operating efficiency.

This method has its shortcomings. Its use in job shops or tool rooms where there are large amounts of handling or setup time is limited because of the inherent difficulty in obtaining cost figures. In analyzing large capital expenditures, there is a tendency to be overly conservative. The chief difficulty lies in obtaining complete and accurate estimates of the information required for the analysis. Few companies have cost records complete enough to supply this summary type of information, which again points up the importance of a cost accounting system that is capable of supplying such information.

The MAPI method is not a substitute for good judgment in making decisions on machinery replacement. However, it is a convenient and practical means of bringing all operating and capital-cost elements to the attention of management. Before the MAPI method can be used, the definition of several items that are used in the worksheet should be obtained. In the following example, the MAPI method is used to solve the replacement problem:

The total operating cost (*B*) for the next year on the new equipment is subtracted from the total operating costs on the old equipment (*A*). Therefore, the total *A* will almost always be larger than the total *B*. Should the reverse be true, the net amount is shown as a negative number. The advantage of the new equipment is a value of $3,000. Dis posal value for the old equipment is most commonly found when considering resale of the equipment. This is usually the value of the equipment in the current used-machine market, which is $5,000.

The present machinery has a loss-of-disposal value of $3,000, which represents the decline in disposal value if the old equipment is kept in service one additional year.

If the old equipment were disposed of or sold, the money could be used as operating capital. As such, it would earn a return; if replacement is postponed, this return will be lost. Use whatever percent you consider proper; 10 percent is suggested as a good average. Furthermore, the total old-equipment cost of $3,500 represents a loss of available capital for the next year. This amount is a capital cost, since it is a charge against keeping the old equipment in service.

Dividing the estimated disposal value by the required investment, the disposal ratio is obtained and found to be 5 percent. Depreciation value is determined from the graph in Exhibit 2. The first step is to locate the service life of the new equipment (eight years) on the horizontal axis. From this point, ascend and locate the disposal ratio (5 percent); then estimate the point at which percentages fall between lines. The percentage value on the vertical line is the depreciation value of the machine (18.5 percent).

If the new equipment is purchased, capital cost represents the capital cost for the next year. The net capital-cost figure is computed by subtracting the total old-equipment capital cost from the total new-equipment capital cost. For instance, the total on line 24 will almost always be larger than the total on line 17 because a large capital cost results from the depreciation chart value and the interest on the investment for one year. Net capital cost is obtained by subtracting the net capital cost of the old equipment from the net operating cost of the new equipment. A plus value indicates that replacement should be made; and the greater the amount, the greater the urgency. A negative value indicates that replacement is premature at this time. A zero or break-even value represents a good investment for replacement, because all the factors that constitute a good machinery investment have been properly considered before arriving at an answer. The advantage would be a minimum loss of savings. This is true because it can

EXHIBIT 2

CHART FOR DETERMINING DEPRECIATION VALUE

Courtesy of Machinery and Allied Products Institute

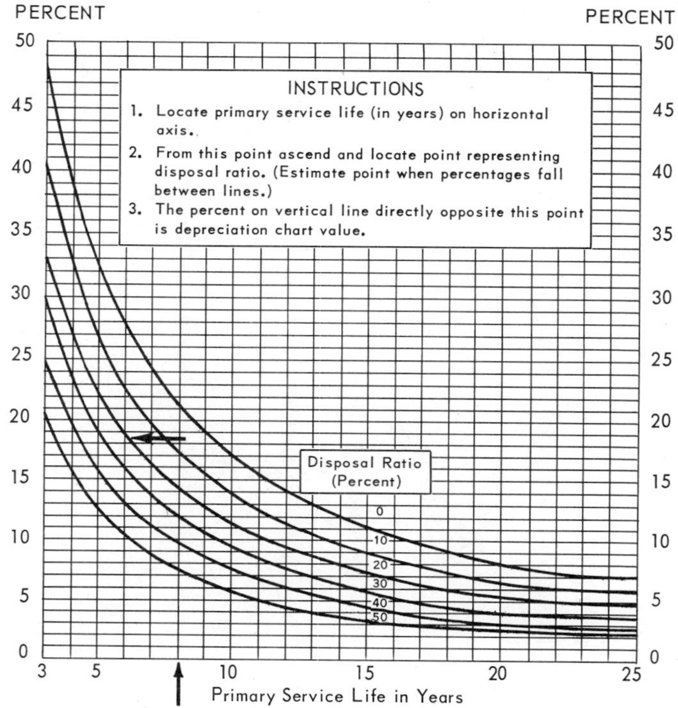

be shown that as operating costs go up, capital costs go down. Therefore, as more years pass, the gap separating the two widens.

Applications for Upper and Middle Management

Determining the capital-investment funds available is a primary stage of the replacement process. This is the function of the company's financial management or its owners and is commonly performed annually when budgets are simultaneously prepared.

The ability to visualize forthcoming technological innovations in raw materials, in manufacturing, and in product design that will be adapted by companies and the industry is important to the industrial environment. The proper selection of new technology by competitors will obviously have a marked impact on a company's welfare. In a few industries technological improvements have come to a virtual standstill, and the problem of predicting future equipment-replacement needs is thus reduced in complexity. But in many industries the changes are rapid, and the task of anticipating the details of the company's technological environment requires continuous research.

One way to avoid glittering generalities about expected alterations in the environment is to anticipate what particular competitors will do about adapting to technological improvements—for example, by learning about the strong points of an improved piece of machinery and competitors' research and development staffs and by observing the types of improvement adopted by the company in the past. The executive is then in a position to make some reasonable conjectures about what type of technological improvements his competitors are likely to consider or adopt. The greatest uncertainty lies with new ideas and discoveries; it is impossible to foresee exactly where the lightning will strike. Yet, even here, there are some guidelines. Executives know that industries are confronted with certain possible alternatives. They know which theories are current, and they can expect solutions eventually, because some extraordinary minds probably are already at work on these problems. But of course they cannot foresee when the solution will be found or exactly what it will be.

Top management must consider more than the basic decision of meeting the payback period or seeing that its return on invested capital is satisfactory. All developments are not made overnight, but certainly top management's failure to take advantage of the rapid improvements in equipment means that it is denying its firm, as well as its consumers and stockholders, the advantages of improved methods of production. We speak of basic productivity of labor as being the output per unit per man-hour, and this is measured, recorded, and published in many sources. However, we need to know as well how the industrial production of any particular firm is doing. If a firm does not have its portion of modern equipment (that equipment which possesses the best-developed productivity criteria), then top management is denying itself the possibility of continued success and growth.

A complete capital-budgeting program involves a number of steps, which can be summarized into procedural and analytical agenda as follows:

1. Prepare a list of investment proposals classified by category: (a) replacement, (b) expansion, (c) cost-saving or product improvement, and

(d) new products. The time horizon covered will include one year, three to five years, and longer-term plans.

2. Study and present probable savings or other justifications for the proposed outlay.
3. Rank proposed projects by a method that reflects interest factors.
4. Develop a rule for applying a cutoff.
5. Determine the total dollar amount of funds required for accepted projects.
6. Determine the amount of internal funds available.
7. Determine the amount of external funds required.
8. Calculate the total of external funds that will be obtained. (This may be less than the total under item 7, and additional rules for rationing funds may have to be formulated.)
9. Decide how the external funds will be raised.
10. Determine the procedures for approval of outlays requested.
11. Determine the procedures for monitoring use of funds appropriated.
12. Evaluate results of expenditures on proposals, projects, and new products.
13. Take corrective actions to improve capital-expenditure policies and actions.

In most business firms proposals for projects exceed the firm's ability or willingness to finance them. Some proposals are good; others are poor. Methods of distinguishing among proposals, therefore, should be developed. Essentially, the end product should rank the proposals as well as delineate the funds required for each of them.

In part, proposals are eliminated because they are mutually exclusive—that is, they are alternate pieces of equipment that perform identical functions. If one piece of equipment is selected to do the job, the others will be unnecessary. Thus, if there is a need to improve the process-control system—for example, in a chemical plant—the work may be performed by additional instruments, by improved instrumentation, or by application of computers. Selection of one method makes it unnecessary to utilize the others.

Independent items, on the other hand, are pieces of capital equipment that are being considered for different kinds of projects or tasks that need to be accomplished. For example, in addition to the items already mentioned, a chemical firm might need equipment to package and measure the end items, which would call for a packaging machine. The purchase of the equipment for this end item would be independent of the equipment purchased for process control. To distinguish among the many items that compete for allocation of capital funds, a ranking procedure must be developed. The ranking procedure involves, first, the calculation of estimated savings from the use of equipment and, second, the translation of the estimated savings into a measure which indicates the advantage of purchasing the equipment.

Factors such as changes in the quality of the project, in the quantity of output, in the quality and quantity of direct labor, in the amount and cost of scrap and rework time, and in maintenance expenses, down time, safety, and flexibility may affect savings on a proposed investment. So many variables are involved that it is obviously impossible to make neat generalizations; yet every capital-equipment expenditure must be examined in detail to avoid additional cost. The difficulty in generalizing, however, should not minimize the crucial importance of a savings analysis. In reality, all of the very elaborate procedures for ranking machines when considering replacement are no better than the basic cost facts that are used as input. All data formulation requires careful judgment in expressing the amount of savings. This is not a routine clerical task to be performed on a mechanical basis; it requires constant monitoring and evaluation of savings estimates by individuals in the firm who are competent to make such evaluations—that is, engineers, economists, cost analysts, and so forth. After the savings estimate has been made, it is utilized for ranking alternative investment opportunities. How this grading is accomplished is of great importance to the management of the firm.

Physical, technological change has not been considered in the ranking process. For the first time some measure of productivity change is available, and this new measure has been incorporated into tables (see Appendix). Management should analyze its cost savings in relation to the actual PCQ change so that the future impact of change will not be so drastic. Without this measure of savings added to the PCQ potential, management could well fit the following description:

> Were it possible to summon from the past a leading business executive of 1850 and send him on a tour of American industry, we may be sure he would greet with astonishment the transformation that has occurred since his day in the arts of management. . . . There is one managerial technique, however, in which our visitor, our executive, would feel much at home. Indeed, it is probably thc only onc hc could takc ovcr without first bringing himself down to date. I refer to the technique of analyzing the replaceability of productive facilities. While the facilities themselves bear little resemblance

> to their prototypes of 1850, the *intellectual process* by which the modern executive arrives at a decision on their replaceability is substantially unchanged. Given the requisite familiarity with present-day equipment, our visitor could at once make analysis as good as the general run. . . .[6]

Available funds having been determined, the final stage of replacement decision making concerns the projects to be undertaken. Projects involving modernization must compete for funds with such requirements as research, advertising, and dividend payments. These evaluation procedures allow the replacement opportunities to be ranked and the degree of urgency compared realistically to other cash outlets. Furthermore, allocations may be made by owners, directors, financial managers, or perhaps a capital-expenditures committee made up of persons from several divisions of the company. The future of the organization is obviously largely determined by the skill with which these allocations are made.

A major pitfall in progressive replacement planning revolves around a misunderstanding of the meaning and intent of accounting allocations. Misconceptions about depreciation, overhead, and profits have distorted business decisions. Whether the old machine is fully depreciated has a slight effect on replacement decisions, for the greatest consideration seems to be that taxes will change by the alterations in depreciable assets.

Profit or business income is the difference between the cost of producing a certain quantity of goods and the revenue produced by selling these goods. Costs of production may be divided into two main categories: capital expenditures and current expenditures. The distinction between these two types of spending is purely one of time, and there is no sharper fundamental distinction between them. Insofar as the benefit accruing from a particular item of expenditure is exhausted during the accounting period in which the expenditure takes place, it is revenue or current expenditure. Insofar as the benefit is prolonged beyond the current accounting period, it is capital or invested expenditure. Therefore, if one takes as a single period a long session of accounting periods, it is obvious that capital and current expenditures will merge and become indistinguishable from each other.

Thus the distinguishing feature of capital expenditure is that it has to be spread over a number of accounting periods. Two main problems result from doing so: The first is determining the period over which the asset will be used in providing services; the second is deciding how to allocate or spread the costs of an asset over the various accounting periods, after being given the single life span.

Applications for Line and Engineering Managements

Alerting management to the most promising replacement opportunity is an important first step in developing applications, since it would be expensive to run a continuous series of studies for replacement of all production equipment. Production foremen should be assigned the responsibility for alerting management to modern opportunities. This procedure works best when the foremen are not so occupied with daily operating problems that they are unable to examine new machines and methods of operation. Another consideration is that some foremen are reluctant to relinquish a familiar machine in exchange for something new. This natural resistance to change must be overcome if production foremen are to be relied upon for suggestions or recommendations for modern equipment. Attendance at machine-tool conventions and visits to other plants, in both the United States and foreign countries, can effectively expose production personnel to the latest trends and help to maintain their interest in the modernizing of operations. Competition among plants is an important motivating factor in replacement programs.

Age, utilization, and manning of machines can probably be checked by a knowledgeable observer without detailed record keeping. Other important factors, such as scrap rates and maintenance costs, usually demand maintenance of careful records. Machines that account for significant quantities of scrap and rework material should therefore be checked for possible replacement or rebuilding.

Another equipment-replacement indicator is the inability to hold tolerances unless speeds and feeds are reduced. Actual running rates should be verified; records are probably necessary to make an accurate comparison between the machine's net acceptable pieces produced per day and its theoretical capacity. Annual maintenance costs for machines are also difficult to estimate without records. Many plants do not yet record costs on an individual-machine basis, and the information, when collected, is often a surprise. Therefore, elimination of costly maintenance is another important justification for machine replacement.

In recent years the problem of equipment replacement has been the subject of close scrutiny in the academic professions and in industry. What has stimulated interest in this problem? The most outstanding and common factor is that industrial management people are becoming more aware of the greater emphasis on and the importance of machine-tool replacement. The total wealth-creating mechanism of our society depends upon the productive ability of machinery and related equipment. For the individual firm this means that capital-expenditure decisions not only form the basic framework for future development, they also can function as major deterrents to its inherent efficiency and competitive ability.

What are some of the problems associated with equipment-replacement studies? The following comments suggest a few: I am about to consider some very important company expenditures. If the decision that I make produces poor results, will I be criticized more than if I had done nothing and allowed operations to function as usual? Of the several good alternatives that I have considered, is there one that I may have overlooked that may equal or even surpass those alternatives that are now before me? What assurance do I have that the many future cost estimates in my possession will remain unchanged? If they are subject to change, have I introduced enough flexibility into the plan to allow for the unforeseen but inevitable changes introduced by new technology? Will the new equipment or procedure actually function more efficiently as claimed? Are the cost savings realistic? What are the possibilities that a new alternative may present itself in a year or so thus making my present decision obsolete and invalid.

To have an effective equipment-replacement program, it is necessary to formulate certain clearly defined policies that will serve as general guides in actual replacement studies. Not only will such policies aid those who are directly involved, they will also make the entire organization aware of the company's deep interest in this area. These policies should encompass such features as the overall attitude toward the operational efficiency of capital assets, the company's expected return on investment, the service-life estimates of various classes of assets, the authority required for approval of such expenditures, the frequency of asset replacement, and the fixing of responsibility regarding analysis of such replacement problems. These are only a few of the problems that must be considered if responsible individuals are to form suitable conclusions for decisions. What is needed, then, is a systematized approach to this entire area.

One important consideration which must not be neglected is the necessity for assembling all the available alternative solutions that will serve as feasible courses of action in a decision-making process. Once the alternatives have been established, the relevant facts may then be considered. Only the differences among alternatives need be considered in making a good replacement study.

The Problem of Overage Machine Tools

The total number of machine tools installed in the nation's metalworking industries has hit an all-time high, surpassing the record set in the Seventh Inventory of 1953. These are highlights that the editors of *American Machinist* and *Metal Machinery* have extracted from thousands of equipment lists, furnished by metalworking plants, that make up the 1963 Ninth Inventory of Metalworking Equipment.

Such cold statistics (64 percent of all machine tools are overage) add up to some dismaying conclusions: The trend to obsolescence that has plagued industry since the end of World War II continues. With such a high percentage of aging, inefficient machines, many metalworking industries may be forfeiting the chance to boost productivity and to remain competitive at home and abroad.

Careful study of the inventory shows that obsolescence is confined to a few types of machine tools. More than 1,300,000 metalcutting machine tools (out of a total of 2,137,000) are at least 10 years old. In each of the 18 major types of metalcutting machines that the inventory lists, the percentage of machines at least 10 years old has risen since the last inventory published in 1958. With minor exceptions, the same pattern is true of metalforming machine tools.

The new inventory shows that the number of machine tools in metalworking plants has reached an all-time high, surpassing the record set by the Seventh Inventory of 1953. Such a record total reverses a trend noted in the 1958 report, which showed a slight drop in machine-tool totals. Compared to the 1958 study, the new *American Machinist* and *Metal Machinery* inventory gives expanded coverage. The most significant expansion is in the Primary-Metals Industry, where plants with some 700,000 wage earners (nearly 10 percent of the total for all of metalworking) have been added.

The Ninth Inventory is based upon returns from 7,370 plants, which employ 27 percent of the wage earners in metalworking. This is by far the largest return in the inventory's history, surpassing 1958's previous high of returns from 5,560 plants. As in every previous inventory, the ten-year yardstick for obsolescence is shown in the latest publication. Such a yardstick sets a uniform standard that applies to all types of equipment and to each industry and area.

Some machines over ten years old are still quite useful;

on the other hand, some machines only five years old can be considered obsolete by present-day standards. The chief advantage of having such a yardstick is that it can be applied repeatedly, inventory after inventory, to show trends in the average life of equipment.

In response to many requests, for the first time the new inventory has collected data on numerical control (N/C). The results of this separate survey will appear soon and provide what is undoubtedly the most complete picture of N/C use to date. Preliminary review of the returns shows that, of the 7,370 plants that responded to the main survey, some 341 have installed one or more numerically controlled machine tools.

Analytical Testing and Inspection

The evaluation of machine-tool deterioration is generally expressed in terms of maintenance cost and down time. These data, when available, provide valuable insight into the trends revealed by a user's experiences. The best present method of estimating the condition of a given machine is by obtaining the subjective judgments of the operator, methods man, foreman, maintenance man, and so forth. Preferable, perhaps, would be specific process-capability statistics from the quality control operations. When speaking of process capability, the entire manufacturing process is included—everything that enters into it—the raw materials, the machine or equipment, the operator's skill, and even the measuring device and the skill of the inspector or tester. Since these factors are subject to fluctuation in value from day to day, it would be advisable to consider a series of process-capability determinations.

Even more important, however, a further insight into productive machine performance may be gained by the analytical testing of a machine tool to determine its capability in terms of tolerances.

The Deterioration Factor

The Productivity Criteria Quotient (PCQ) is a measure only of the obsolescence factor of relative machine-tool performance. To demonstrate the total advantage of a new machine tool, the equally important factor of deterioration must be considered. What actually happens, of course, is that from the day it is installed, a new machine tool begins to deteriorate in relation to its original capability. This process is gradual in most cases, as our depreciation policies acknowledge; but by the end of any substantial period deterioration is noticeable.

Standard capacity is usually defined as two-shift operation; however, many departments vary, and actual shift utilization must be considered. Three-shift operation would deteriorate a machine tool in six years—the equivalent of nine years of two-shift operation. Preventive maintenance can only arrest—not avoid—deterioration. The complex design of today's machine tools induces more rapid obsolescence and deterioration. Better machine design concepts, such as vibration control, improvements in metallurgy, and closer construction tolerances, are offset by higher utilization.

The machines work harder more of the time. Translated into actual loss of productivity, the measurement of this deterioration is a natural extension of the PCQ concept. This area is currently under investigation to develop a better way of measuring—and estimating—the total effect of deterioration. To do less is to estimate a deterioration value of zero, which is, of course, incorrect. Exhibit 3 illustrates this concept.

Which factor is greater in terms of justifying machine-tool replacement—obsolescence or deterioration? That's a knotty question! Certainly we must consider all elements of both factors to know the total advantage of a new machine tool.

The Army Materiel Command, represented by the U.S. Army Production Equipment Agency (known as PEQUA) at the Rock Island Arsenal, has done considerable work in developing measurements of the productive capability of machine tools to evaluate equipment-replacement requests. This work is based on the PCQ concept of measuring obsolescence and has, through analysis of hundreds of replacement requests, substantiated PCQ theory to a very satisfactory degree.

For a number of years, machine-tool replacement analysis[7] has incorporated a term called PIR—Productivity Increase Ratio. PIR reflects the ratio in improved operating efficiency of the "proposed equipment" to that of the "present equipment." This ratio is normally developed through engineering studies and through vendors' estimates of production potential. PEQUA has developed PIR tables[8] to eliminate engineering studies of individual parts to be processed. These tables are, as mentioned before, based entirely on the PCQ concept, but contain additional factors, such as average service life (estimated at thirteen years), fair wear and tear (depreciation), and a desired payback period. A five-year payback is acceptable to the Army; but note that this is equivalent to a ten-year cash flow, since it does not pay corporate taxes.

The PEQUA group has also developed another set of tables by machine type to suggest which type of machine-

EXHIBIT 3

OBSOLESCENCE AND DETERIORATION

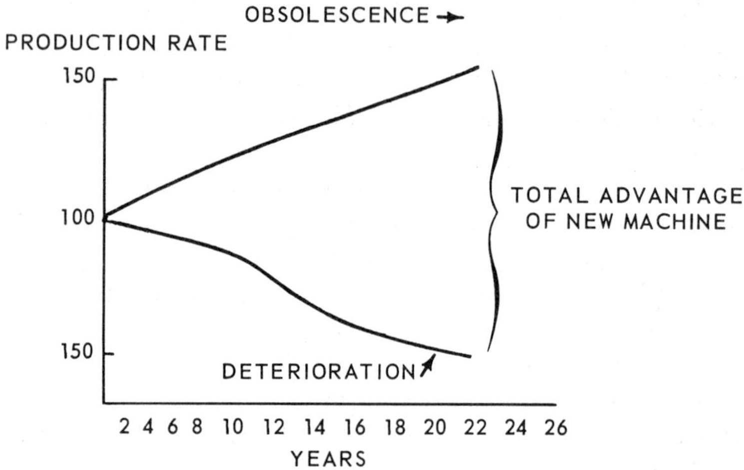

tool might offer the greater potential return. This was a measure of the relative improvements among different classes of machines and was based on calculating the percent change in PCQ per year by machine type. Certain machine types have more design changes that directly influence productive ability than other types of machines. This, of course, would be logical. Everything else being equal, this guide suggests that the category of machine-tool type offers a greater potential return in an equipment-modernization program.

Other potential applications of the PCQ concepts include the development of a computer program for the systematic and automatic identification of machine tools as candidates for replacement. Think of the budget and forecasting help this would provide! It's coming, too, for this certainly is the age of systems engineering. With the further development of the PIR tables, such a program could well prepare the basic justification data.

PCQ information can be employed as one consideration in measuring the relative merits of farm-out sources. Certainly, using this concept as a sales tool can be visualized by machine-tool vendors or by control manufacturers. It can also be helpful when acquiring surplus machine tools, since the years in which major innovations were introduced can be easily determined.

In Conclusion

Let us summarize the points we have made:

- New and improved equipment is now available for company growth. The technology explosion makes more rapid replacement policies mandatory in order to keep pace with the problem of rapid obsolescence.
- The replacement function must be analyzed and measured more closely than ever before to be able to plan for the future, and it must be vigorously supported in investment budgets and policies.
- Manufacturing services should be supported in the development, extension, and application of PCQ. Toward that end, a complete set of PCQ curves and forecasts for the 13 basic machine-tool types have been prepared (see Appendix).

Notes

[1] Lawrence L. Bethel, Franklin S. Atwater, George H. E. Smith, and Harvey A. Stackman Jr., *Industrial Organization and Management,* 3rd ed. (New York: McGraw-Hill Book Company 1956), pp. 351–6.

[2] C. Canby Balderston, Victor I. Karabasz, Robert J. Brecht, and Robert J. Riddle, *Management of an Enterprise* (New York: Prentice-Hall, Inc., 1949), pp. 88–94.

[3] Manley H. Jones, *Executive Decision Making,* rev. ed. (Homewood, Illinois: Richard D. Irwin, Inc., 1962).

[4] *Business Week,* January 2, 1949, p. 56.

[5] Eugene L. Grant and W. Ireson, *Principles of Engineering Economy* (New York: Ronald Press Co., 1950).

[6] George Terborgh, Research Director, Machinery and Allied Products Institute, *An Introduction to Business Investment Analysis,* p. 5.

[7] *Plant-Equipment Modernization Program—Regulation No. 700–22, Procurement, AMC Production Base Support Program—Regulation No. 715–33,* U.S. Army Materiel Command, Superintendent of Documents, U.S. Government Printing Office, Washington, D.C.

[8] *A.S.T.M.E. Technical Paper, SP66–09,* and S. H. Miller, *Standardization of the Productivity Increase Ratio,* U.S. Army Production Equipment Agency (PEQUA), U.S. Army Materiel Command, Rock Island Arsenal, Illinois.

Appendix

All exhibits in this Appendix
courtesy of *American Machinist,* October 7, 1968.

PCQ ratings for 13 types of machines, 1955-1967

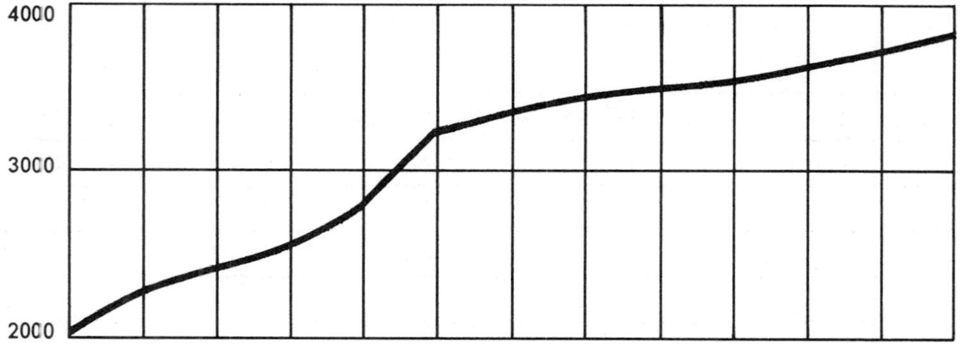

Machine types	'55	'56	'57	'58	'59	'60	'61	'62	'63	'64	'65	'66	'67
Cutoff saws	32	50	64	72	86	96	99	99	102	105	122	137	137
Contour saws	77	77	77	90	90	93	93	93	93	103	126	131	136
Cylindrical grinders	180	203	206	229	249	274	274	274	274	274	284	289	289
Surface grinders	80	80	107	115	118	151	151	153	166	183	197	205	225
Vertical turret lathes	405	435	470	505	630	640	655	670	690	690	705	711	711
Turret lathes	312	345	345	375	385	468	481	481	494	509	522	528	530
Engine lathes	280	286	287	290	309	392	392	396	396	409	418	424	427
Jig borers	124	137	170	173	196	253	253	253	253	270	287	295	295
Horizontal boring, milling	276	300	350	364	382	472	482	494	494	502	514	514	529
Milling machines	148	156	166	166	172	196	198	198	198	201	208	214	224
EDM machines	10	13	30	30	30	45	45	47	55	63	71	77	110
Sensitive drills	22	26	32	48	52	70	70	78	83	96	103	108	121
Radial drills	68	76	84	104	117	167	167	167	167	167	175	175	185
Total for 13 types	2014	2184	2338	2561	2816	3317	3360	3403	3465	3572	3732	3808	3919

Cut-off saws

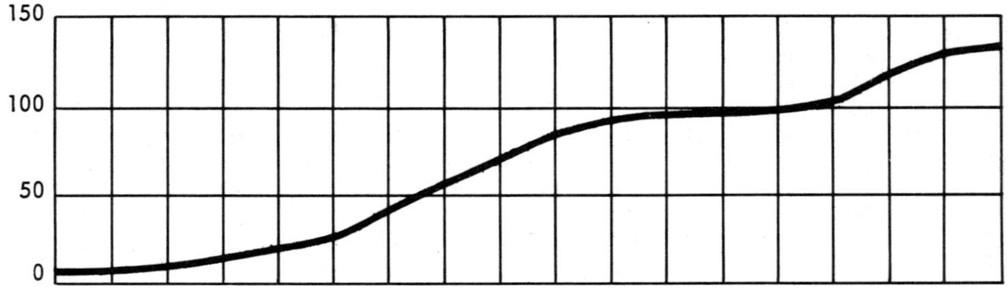

Improved features	Wt.	'50	'51	'52	'53	'54	'55	'56	'57	'58	'59	'60	'61	'62	'63	'64	'65	'66	'67
Automatic control	3										6						3	3	
Feed mechanism	2				2			6	2	2		2							
Straightness of cut	2				2			2	2	2									
Other features	1				3	3		1	3		1	3	1		1	1	2		
Total		0	0	0	7	3	0	9	7	4	7	5	1	0	1	1	5	3	0
PCQ		12	12	12	26	32	32	50	64	72	86	96	99	99	102	105	122	137	137

Contour saws

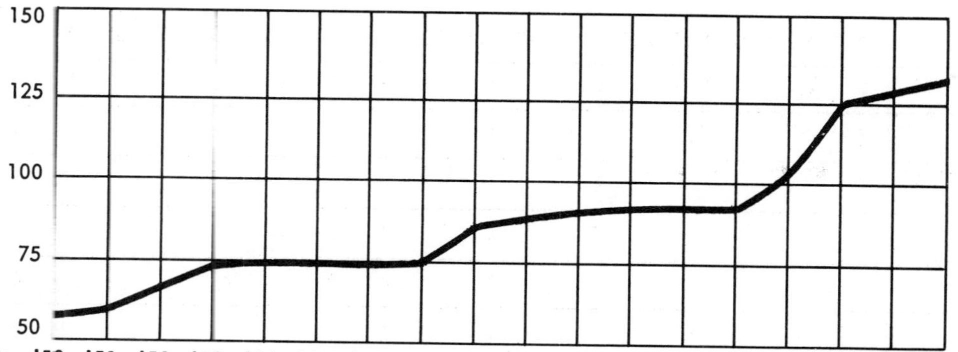

Improved features	Wt.	'50	'51	'52	'53	'54	'55	'56	'57	'58	'59	'60	'61	'62	'63	'64	'65	'66	'67
Feeding mechanism	2			2						2							2		
Speeds	2																		
Horsepower	1									1						1	1		
Band size	1				1					1									
Straightness of cut	1															1	1		
Other features	1	1			1							1				1	3	1	1
Total		1	0	2	2	0	0	0	0	4	0	1	0	0	0	3	7	1	1
PCQ		63	63	70	77	77	77	77	77	90	90	93	93	93	93	103	126	131	136

Cylindrical grinders

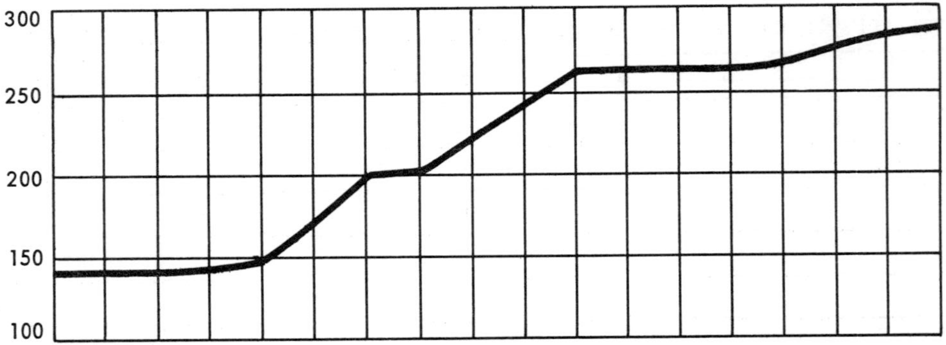

Improved features	Wt.	'50	'51	'52	'53	'54	'55	'56	'57	'58	'59	'60	'61	'62	'63	'64	'65	'66	'67
Work head	2	4					2	2		4	4	4							
Wheel head	2				2		6	4		2	4	2							
Wheel feeds	2					2	2	2		2		2							
Table speeds	2											2						2	
Other features	1							1	1	1							3		
Total		4			2	2	10	9	1	9	8	10	0	0	0	0	3	2	0
PCQ		145	145	145	150	155	180	203	206	229	249	274	274	274	274	274	284	289	289

Surface grinders

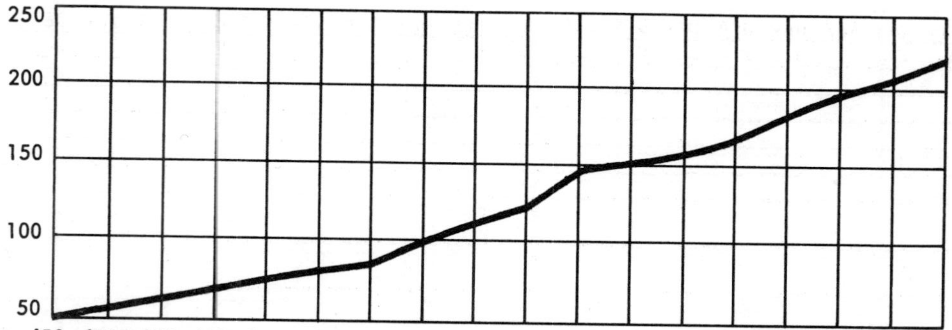

Improved features	Wt.	'50	'51	'52	'53	'54	'55	'56	'57	'58	'59	'60	'61	'62	'63	'64	'65	'66	'67
Automatic controls	3															3	3		3
Wheel head	2		4	2					6	2	2	6			2	2	2	2	
Work table and head	2	2			2	4			6			4			6	2		2	2
Table speeds	2								2	2		4				2	2		
Horsepower	1				1	1	1			1		1		1		1			
Other features	1	2	1	1		1			1			4			1	1	2		1
Lubrication system	1								1			1				1	1		
Total		4	5	3	3	6	1	0	16	5	2	20	0	1	9	12	10	4	6
PCQ		50	58	63	68	78	80	80	107	115	118	151	151	153	166	183	197	205	225

Vertical turret lathes

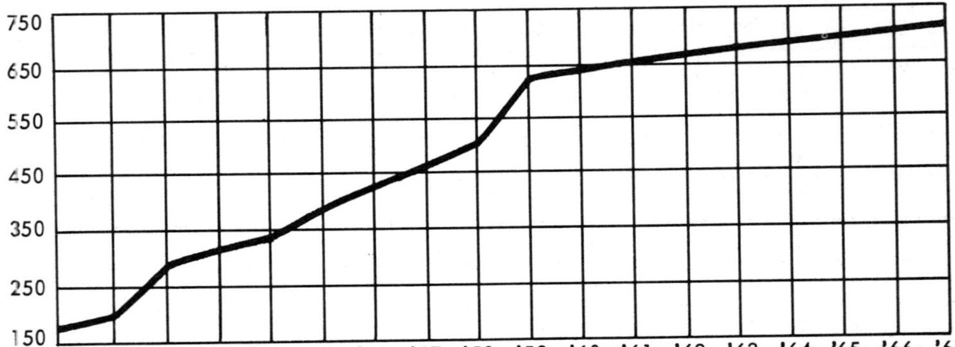

Improved features	Wt.	'50	'51	'52	'53	'54	'55	'56	'57	'58	'59	'60	'61	'62	'63	'64	'65	'66	'67
Automatic controls	3								3	3	3		3	3	2		3		
Bed construction	2			4			2				2								
Head, cross-rail construction	2			10	2		4	2	4		8	2			2				
Headstock gear shifting	2			2			6				4							2	
Feeds, feed mechanism	2	2		2				4		4	4								
Table speeds	1			1							1								
Table bearings	1		1		3													1	
Horsepower	1	1			1						1								
Clutches	1			1			2				1								
Lubrication system	1			1		1					1								
Total		3	1	21	6	1	14	6	7	7	25	2	3	3	4	0	3	3	0
PCQ		190	195	300	330	335	405	435	470	505	630	640	655	670	690	690	705	711	711

Turret lathes

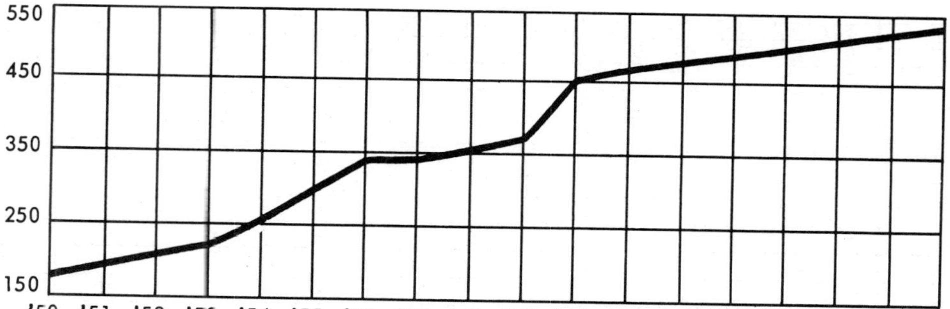

Improved features	Wt.	'50	'51	'52	'53	'54	'55	'56	'57	'58	'59	'60	'61	'62	'63	'64	'65	'66	'67
Automatic controls	3									3		15					3		
Headstock speeds	2	4				4	2	4			4					2	2	2	
Less spindle run out	2	2					4			2		2				2			
Feeds	2	2					4			2		4			2	2			
Bed and way	2		2			2	6	4				4	2		2				
Carriage	1		1		4	1	2	1		1		3			1			1	
Horsepower	1		2		1	1	2	2				2							
Headstock gear shifting	1		1			2	2			1		1	1						1
Lubrication system	1		1			1	1	1											
Clutches	1	1	1	2				1		3		2	2						
Total		9	8	2	5	11	23	13	0	12	4	33	5	0	5	6	5	3	1
PCQ		183	208	213	226	254	312	345	345	375	385	468	481	481	494	509	522	528	530

Engine lathes

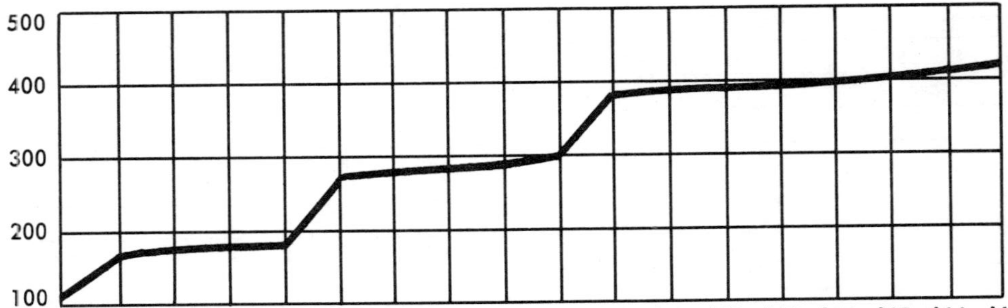

Improved features	Wt.	'50	'51	'52	'53	'54	'55	'56	'57	'58	'59	'60	'61	'62	'63	'64	'65	'66	'67
Automatic controls	3											12		3		4		3	
Tracer heads	2	10					2					4							
Headstock speeds	2	2		2	2		12				2	4							
Spindle bearings, construction	2						6				2	8				2			
Feeds	2		2			4	4				2	6							
Bed and way	2	4					12	2			4	10				2			
Carriage	1				2		11	1	1	1	1	6					2		
Horsepower	1	1		1			8				1	4							
Headstock gear shifting	1						3	1			2	4					1		1
Lubrication system	1	2					5	1		1	1	2							
Clutches	1			1	1		5					4				1			
Additional features	1						5					2					3		
Total		19	2	4	5	4	73	5	1	2	15	66	0	3	0	9	6	3	1
PCQ		170	173	178	184	189	280	286	287	290	309	392	392	396	396	409	418	424	427

Jig borers

Improved features	Wt.	'50	'51	'52	'53	'54	'55	'56	'57	'58	'59	'60	'61	'62	'63	'64	'65	'66	'67
Automatic controls	3							3	3		6	6				3	3		
Ability to hold tolerances	3								3									3	
Feeds	2								2			4							
Speeds	2								2			4					2		
Spindle construction	2						2									2			
Horsepower	1						1					1							
Bed and way	1						2			1	1	2							
Clutches	1						2												
Lubrication system	1						2	1											
Total		0	0	0	0	0	9	4	10	1	7	17	0	0	0	5	5	3	0
PCQ		94	94	94	94	94	124	137	170	173	196	253	253	253	253	270	287	295	295

Horizontal boring and milling machines

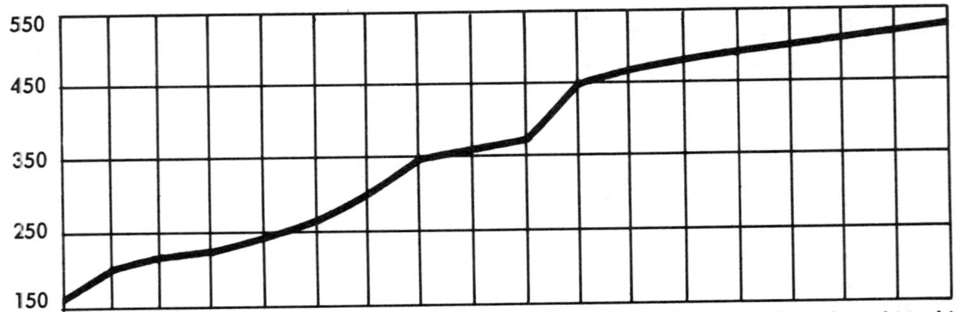

Improved features	Wt.	'50	'51	'52	'53	'54	'55	'56	'57	'58	'59	'60	'61	'62	'63	'64	'65	'66	'67
Automatic controls	3							3	3		3	6	3	3			3		
Feeds and speeds	2	6	6			2	2	2	2			14	2			2	2		
Bed and way	2	4	2	2		2	2	2	2	2	2	4				2			2
Column design	2	2	2			2	4	4	4		2	6							
Clutches	2	8	4			2	2		4	4	2	2		2					
Less spindle drift	1		2				1	1	3			2							
Horsepower	1	4	2			2	2		2			3		1					
Ability to hold tolerances	1	1	1						2			1					1		
Additional features	1	1	3		1		4		3	1		7							1
Total		26	22	2	1	10	17	12	25	7	9	45	5	6	0	4	6	0	
PCQ		172	216	220	222	242	276	300	350	364	382	472	482	494	494	502	514	514	529

Milling machines

Improved features	Wt.	'50	'51	'52	'53	'54	'55	'56	'57	'58	'59	'60	'61	'62	'63	'64	'65	'66	'67
Feeds and speeds	2	2					8	2	4			2					4		
Clutches	2						4					2				2			2
Knee, saddle construction	2						16				2	4						2	
Horsepower	1	1					4	1	1			1							
Additional features	1						3	1			1	3	1					1	1
Total		3	0	0	0	0	35	4	5	0	3	12	1	0	0	2	4	3	3
PCQ		78	78	78	78	78	148	156	166	166	172	196	198	198	198	201	208	214	224

Electrical-discharge (EDM) machines

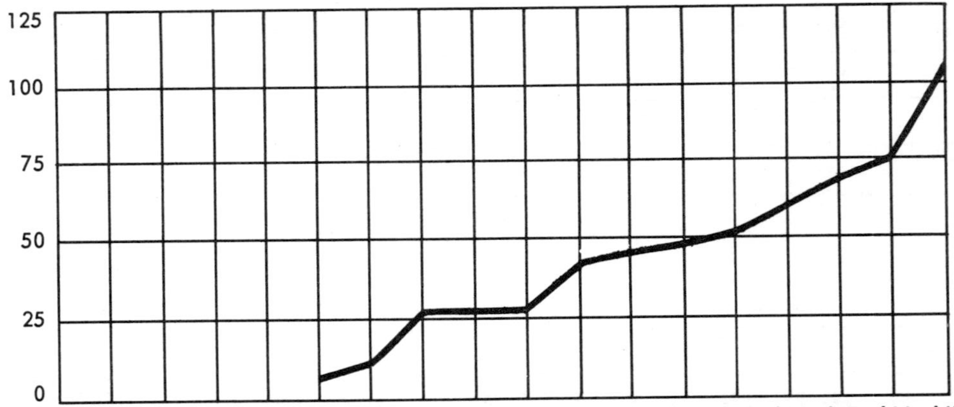

Improved features	Wt.	'50	'51	'52	'53	'54	'55	'56	'57	'58	'59	'60	'61	'62	'63	'64	'65	'66	'67
Rate (amperes)	3								3							3	3		6
Frequency	1											1						1	
Electronic envelope (overcut)	2								2			2							4
Coolant	1							1						1					
Electrode material	3						3					3			3				
Additional features	1														1	1	1	2	3
Total							3	1	5	0	0	6	0	1	4	4	4	3	13
PCQ		– data not available –					10	13	30	30	30	45	45	47	55	63	71	77	110

Sensitive drills

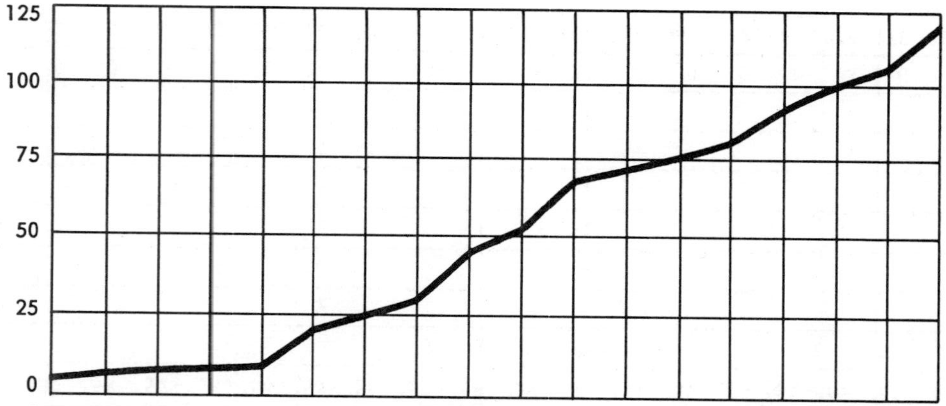

Improved features	Wt.	'50	'51	'52	'53	'54	'55	'56	'57	'58	'59	'60	'61	'62	'63	'64	'65	'66	'67
Automatic controls	3									3		6		5	3	5			3
Spindle construction	2						2	2		4	2						2		2
Horsepower	2						2												
General features	1	1		1			2		3	1		3				3	2	2	
Total		1	0	1	0	0	6	2	3	8	2	9	0	5	3	8	4	2	5
PCQ		8	8	10	10	10	22	26	32	48	52	70	70	78	83	96	103	108	121

Radial drills

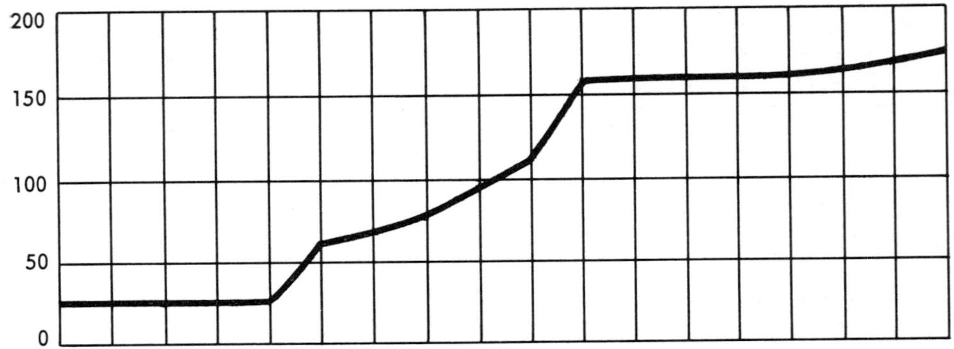

Improved features	Wt.	'50	'51	'52	'53	'54	'55	'56	'57	'58	'59	'60	'61	'62	'63	'64	'65	'66	'67
Automatic controls	3								3	6	3	12							3
Feeds	2						2												
Speeds	2						6					4							
Gear shifting	1						2	2		1		1							
Clutches	1						1	1				2					1		1
Horsepower	1						1										1		
General features	1						2			1	2	1					1		
Total		0	0	0	0	0	14	3	3	8	5	20	0	0	0	0	3	0	4
PCQ		33	33	33	33	33	68	76	84	104	117	167	167	167	167	167	175	175	185

About the Author

Lawrence C. Hackamack heads the Department of Management, College of Business, Northern Illinois University. He received his M.S. degree in business from Western Illinois University and his Ph.D. degree in management from the State University of Iowa. Formerly the National Educational Director for the American Production and Inventory Control Society, Dr. Hackamack is the originator of the Productivity Criteria Quotient, and has been active as a teacher and management consultant for many years.

$4.50
BUSINESS

Other publications from the AMA Management Bookshelf

Three-Dimensional Foremanship — *Frank A. Busse*

A portrait of the foreman in his trilateral role of supervisor, manager, and instructor—useful as either a textbook for the foreman or a guide for management. Emphasizes the importance of his job by means of a unique chart that breaks down his duties into 31 categories and 6 major classifications.
$9.75/AMA members: $6.50

Objectives and Standards of Performance in Production Management — *Ernest C. Miller*

Establishes a solid foundation upon which companies can build effective programs for planning and controlling production performance. Stresses the value of integrating job targets with company goals, describes the development of results programs in major manufacturing firms, and examines actual objectives and standards set for both line and staff positions in specific companies.
$12.00/AMA members: $8.00

A Management Role for Quality Control — *John T. Hagan*

Describes a "new breed" philosophy that puts quality control in its proper context—as management's strong right arm. Identifies and explores all aspects of the quality-control function, and shows how it can be organized, planned, and executed in the best interests of the company without compromising its contribution to overall profits.
$9.00/AMA members: $6.00

American Management Association, Inc.
135 West 50th Street New York, N.Y. 10020

Cover design by Vincent Scotti
8144-2131-8